Explaining Renewable Energy

This undergraduate text aimed primarily at high schoolers and lower-level undergraduates focuses on explaining how the various forms of renewable energy work and the current ongoing research. It includes sections on non-scientific aspects that should be considered such as the availability of resources. A final chapter covers methods of removing carbon dioxide from the atmosphere.

Renewable energy is currently on everyone's mind in the context of climate change. This text provides students with an introduction to the science behind the various types of renewable energy, enabling them to access review literature in the field and options that should be considered when selecting methods.

Explaining Renewable Energy

Authored By
Elaine A. Moore

CRC Press
Taylor & Francis Group
Boca Raton London New York

CRC Press is an imprint of the
Taylor & Francis Group, an **informa** business

First edition published 2023
by CRC Press
6000 Broken Sound Parkway NW, Suite 300, Boca Raton, FL 33487-2742

and by CRC Press
4 Park Square, Milton Park, Abingdon, Oxon, OX14 4RN

Library of Congress Cataloging-in-Publication Data

Names: Moore, Elaine (Elaine A.), author.
Title: Explaining renewable energy / authored by Elaine A. Moore, Honorary Associate, The Open University, UK.
Description: First edition. | Boca Raton : CRC Press, [2023] | Includes bibliographical references and index. |
Identifiers: LCCN 2022027006 (print) | LCCN 2022027007 (ebook) | ISBN 9781032278414 (hardback) | ISBN 9781032275758 (paperback) | ISBN 9781003294337 (ebook)
Subjects: LCSH: Renewable energy sources. | Carbon dioxide mitigation.
Classification: LCC TJ808 .M66 2023 (print) | LCC TJ808 (ebook) | DDC 333.79/4--dc23/eng/20221003
LC record available at https://lccn.loc.gov/2022027006
LC ebook record available at https://lccn.loc.gov/2022027007

ISBN: 978-1-032-27841-4 (hbk)
ISBN: 978-1-032-27575-8 (pbk)
ISBN: 978-1-003-29433-7 (ebk)

DOI: 10.1201/9781003294337

Typeset in Bembo
by Deanta Global Publishing Services, Chennai, India

Contents

Preface

Global warming has already led to increases in wild fires, floods, droughts and other extreme weather events.

It has been known for over 25 years that increasing concentrations of greenhouse gases in the atmosphere due to human activity are the main cause of global warming. In 1996, I was part of a team producing a Science Foundation Couse for the Open University. One book in this course (written by another academic) showed graphs of the increase in atmospheric temperature during the 20th century. Possible explanations were explored, and it was shown that the proposed alternatives to human activity as a cause were insufficient to explain the observed rise in temperature.

As a solid state chemist, I have observed that conferences in this area in recent years are often dominated by research into methods of reducing atmospheric carbon dioxide, particularly batteries and carbon capture.

I was pleased therefore to be asked to write this book. However, this only covers some of the problems that need to be tackled. For example, energy use, particularly by inhabitants of 'rich' counties, needs to be reduced.

I would like to thank Jack Drever who read and commented on the chapter on wind power. I would also like to thank the reviewer of the first draft, who provided useful comments on the text as a whole.

I also owe thanks to the editorial team at Taylor & Francis, in particular Hilary Lafoe, who suggested the project and offered support throughout the production process.

Acknowledgement

I wish to acknowledge the use of the Chemical Database Service at Daresbury, Cheshire, England.

(Fletcher, D. A., Meeking, R. F., and Parkin, D., 'The United Kingdom Chemical Database Service', *J. Chem. Inf. Comput. Sci.* **36**, 746–749, 1996), and the Inorganic Crystal Structure Database.

(Berghoff, G. and Brown, I. D., 'The Inorganic Crystal Structure Database (ICSD)', in *Crystallographic Databases*, F. H. Allen et al., eds. Chester, International Union of Crystallography, 1987.) Also WebLab Viewer Pro or Discovery Studio Visualiser from Dassault Systèmes Biovia Corp.

Author

Elaine A. Moore studied chemistry as an undergraduate at Oxford University, England, and then stayed on to complete a DPhil in theoretical chemistry with Peter Atkins. After a two-year postdoctoral position at the University of Southampton, England, she joined the Open University, UK (OU), in 1975, becoming a lecturer in chemistry in 1977, senior lecturer in 1998 and reader in 2004. She retired in 2017 and currently has an honorary position at the Open University.

She has produced OU teaching texts in chemistry for courses at levels 1, 2 and 3 and written texts in astronomy at level 2 and physics at level 3. She is a co-author of *Metals and Life* and of *Concepts in Transition Metal Chemistry*, which were part of a level 3 Open University course in inorganic chemistry, and were co-published with the Royal Society of Chemistry. She was team leader for the production and presentation of an Open University level 2 chemistry module delivered entirely online. She is a Fellow of the Royal Society of Chemistry and a Senior Fellow of the Higher Education Academy. She was a co-chair for the successful departmental submission of an Athena Swan bronze award.

Her research interests are in theoretical chemistry applied mainly to solid-state systems, and she is the author or

co-author of more than 50 papers in refereed scientific jour-
nals. A long-standing collaboration in this area led to her
being invited to help run a series of postgraduate workshops
on computational Materials Science hosted by the University of
Khartoum, Sudan.

Chapter 1

Introduction

Increasing concentrations of greenhouse gases in the atmosphere are the main cause of global warming. This has already led to increases in wild fires, floods and other extreme weather events. The name greenhouse gas comes from a process that was once thought to be responsible for heating greenhouses, but this is no longer thought to be so. Greenhouse gases do, however, heat up our atmosphere. Solar radiation is absorbed by the surface of the Earth, which then radiates infrared radiation into the atmosphere. Greenhouse gases capture this radiation and then re-emit it. This warms the lower atmosphere. Common greenhouse gases are carbon dioxide, methane and water vapour. Pollutant gases such as ozone and nitrogen oxides are also greenhouse gases. Nitrogen, N_2, which makes up the bulk of our atmosphere, is not a greenhouse gas, and neither is oxygen, O_2. The concentration of water vapour depends on the atmospheric temperature rather than the amount emitted, so the primary targets for emission reduction are carbon dioxide and methane.

Fossil fuels (coal, oil, gas) react to give carbon dioxide when burned. Thus, one way to reduce the emission of carbon dioxide is to change from fossil fuel–powered energy sources to ways of energy production that do not emit

DOI: 10.1201/9781003294337-1

greenhouse gases. Such energy sources are labelled alternative energy or clean energy. Those that use resources that are continually replenished such as sunshine, wind, rivers, waves, tides and heat within the earth are known as renewable energy sources. Not all clean energy sources are renewable sources. While nuclear energy power stations do not emit carbon dioxide, uranium ores are not continually being replenished. In addition, they produce radioactive waste, which has to be contained.

This book concentrates on renewable energy. You are going to be introduced to how renewable sources generate energy. I will also cover the materials needed to build the structures necessary for each and drawbacks such as intermittent supply and environmental impacts. In considering the environmental impacts of renewable energy sources, it is worth bearing in mind that fossil fuel energy sources also have an impact on the environment and in addition they continuously produce pollutants such as nitrogen oxides and small particles that can enter your lungs.

In addition to strictly renewable energy systems, I look at using hydrogen gas as an alternative to fossil fuels. Hydrogen only produces water when burned. However, there are no reserves of hydrogen, and it has to be manufactured. Using a renewable energy source to provide electricity to make hydrogen gas yields the so-called green hydrogen.

Some renewable energy sources only produce intermittent energy, so energy storage, for example in batteries, can be important.

The last chapter covers carbon capture, storage and reaction of carbon dioxide to produce useful products. The aim here is to provide additional reduction of the carbon dioxide concentration by removing it from the atmosphere.

Methane emissions also need to be reduced, but this is not covered here.

I have not covered economic considerations. The position here is continually changing. The fluctuating price of fossil fuel prices is one factor. The wider spread of renewable energy systems, particularly wind turbines and solar cells, leading to a drop in the price of the energy they produce, is another factor.

Finally using less energy will also be needed to reduce global warming.

Chapter 2

Solar Energy

Here we consider two ways of using energy from the Sun. First by using sunlight to produce electricity via photovoltaic cells and second by heating water for homes, thermal solar cells. We discuss how photovoltaic cells work and the different types that are in use or proposed. The materials needed to produce these cells and their advantages and disadvantages are also covered.

2.1 Photovoltaic (Solar) Cells

Most solar cells in use today are based on silicon doped with very small amounts of other elements. We begin by introducing the science behind such cells.

2.1.1 Semiconductors

Semiconductors conduct electricity but not as well as metals. In addition, their conductivity rises with temperature, whereas that of metals decreases with temperature. To see why, we shall use silicon as an example. A silicon atom has four valence electrons, two 3s electrons and two 3p electrons.

DOI: 10.1201/9781003294337-2

In the atom, the s and p electrons occupy distinct energy levels. In a solid of N atoms, the s and p orbitals form two bands of energy levels. Within each band, the levels are close together so that they can be approximated as a continuous band of energies. Figure 2.1 shows these bands. The lower energy band is known as the valence band and the upper as the conduction band. At 0 K the valence electrons only occupy the lower band. To conduct electricity, vacancies must occur in one or both bands. This can be achieved by heating the solid or by shining light on it to promote electrons from the valence band to the conduction band.

Conduction can also be increased by doping the silicon with atoms of other elements that have more or fewer valence electrons. Common dopants are phosphorus with five valence electrons and boron with three. Dopants with more valence electrons form donor levels below the conduction band. The electrons in these levels are easily promoted to the conduction band. Semiconductors with these dopants are n-type

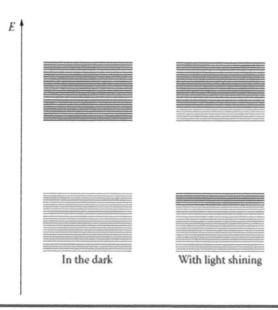

Figure 2.1   **The promotion of electrons from the valence band to the conduction band by light.**

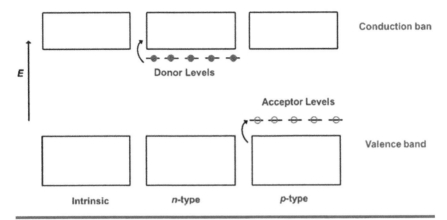

Figure 2.2 **Intrinsic, n-type and p-type semiconductors showing electrons in the conduction band (negative charge carriers) and energy levels to which electrons from the p-type are promoted to leave vacancies in the valence band (positive holes).**

semiconductors. Dopants with fewer valence electrons give rise to vacancies in the valence band. Semiconductors with these dopants are p-type. Figure 2.2 compares the energy levels of Intrinsic (non-doped) semiconductors, n-type and p-type semiconductors.

2.1.2 The p–n Junction

Single crystal silicon solar cells contain adjacent regions of n-type doped silicon and p-type doped silicon. At the junction of these two regions (the p–n junction), electrons move from the n-type region to the p-type region, making the n-type region positively charged and the p-type negatively charged. The flow is opposed by having to move electrons into a negatively charged region and an equilibrium is set up. When light is shone onto the p–n junction, electrons in the valence band at the junction absorb the light and are promoted to the conduction band, where they move into the n-type region. If the cell is made part of an electrical circuit, these electrons will travel through the n-type region and into the circuit, returning into the p-type region. Figure 2.3 illustrates this process.

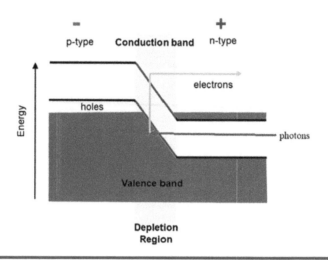

Figure 2.3 The effect of light on a p–n junction.

2.1.3 *Materials Required*

Silicon is a shiny silver material. The shine is due to reflected light. Therefore, to increase the efficiency of silicon-based solar cells, they are coated with an antireflection material such as sulfur nitride or indium tin oxide. Light reflected from the surface of the coating is out of phase with that from the silicon, leading to destructive interference. This increases the percentage of sunlight reaching the p–n junction. It is this antireflective layer that gives the cell its blue colour. Solar cells also need a coating to keep them clean and protect them from rain, and metal contacts to carry the current

Figure 2.4 shows the composition of a typical silicon solar cell module. The silicon layer is attached to conducting layers on top and underneath. Above the top conducting layer is a layer of antireflective coating and a cover of glass. Below the bottom conducting layer is a sheet of polymer, polyvinyl fluoride (PVF) and attached to this is a metallic contact attached to wires to distribute the electricity produced. Ethylene vinyl acetate layers are placed between the top conducting layer and the antireflective coating and between the bottom conducting

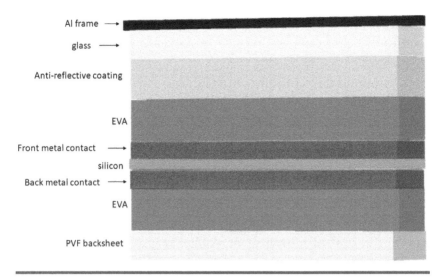

Figure 2.4   **Composition of a typical single crystal silicon solar cell. EVA = ethylene vinyl acetate.**

layer and the PVF sheet. These layers protect the layers between them from substances that may affect the working of the cell such as dust, moisture and ice. The module is stabilised by a frame of aluminium and the aluminium alloy, $AlMg_3$.

Single silicon crystal cells require very pure silicon. Even the dopants (P and B) should only be present at a concentration of $1:10^{10}$ Si atoms. Although silicon is a very abundant element, it is usually found in combination with oxygen, and the major raw material for single crystal silicon cells is quartz sand. This is quarried and transported to the factory. Mining and transport require energy and currently lead to carbon dioxide emissions. The oxide (or other mineral) first needs to be converted to trichlorosilane, $SiHCl_3$, which is then distilled and decomposed at 1000°C. This is a process using large amounts of energy. A 2011 paper (V. M. Fthenakis and A, C, Kim (2011) *Solar Energy*, **85**, 1609–1628) estimates this as 45% of the total energy required to make a crystalline silicon cell module. The process is well established for devices using silicon chips. It is possible to use scrap silicon from the manu-facture of these devices and use of recycled silicon could be considered.

The metals carrying the current away from the cell are easily removed and can be recycled. Aluminium is needed in the greatest quantity as it is used for the module frame. Aluminium is plentiful and can be recycled. Copper and silver are good electrical conductors. Copper is widely used in electrical circuits and pipes for gas and water and its supply has become limited. There is a serious threat to silver supplies in the next 100 years. Recycling is particularly useful for silver due to the cost of mined silver. Nickel is also under threat of limited supply. Iron is abundant.

Other materials are glass, ethylene vinyl acetate, titanium dioxide, polyvinyl fluoride and the antireflective coating. Glass is recycled. Titanium is abundant. For an antireflective coating of indium tin oxide, indium is one of the least available elements and is obtained as a by-product of zinc mining. It is toxic if more than a few milligrams are consumed. Tin is of limited availability and is obtained from conflict minerals.

Installation includes manufacture of an aluminium frame, mounting structures, cabling and the electronics required to link the cells to appliances or the National Grid.

At the end of their life the cells will need to be removed from the roof (or ground for solar cell farms) and transported to where they will be disposed of or recycled.

2.1.4 Other Photovoltaic Cells

Alternatives to single crystal silicon include polycrystalline silicon, amorphous silicon, conducting organic polymers, cadmium telluride, gallium arsenide and hybrid perovskites (compounds of formula AMX_3, where A is a substituted ammonium ion such as $N(CH_3)_3H$, M is tin or lead and X is a halogen). **Polycrystalline and amorphous silicon** are cheaper to produce, although less efficient than single crystal silicon.

Conducting organic polymers are cheaper to produce than high purity silicon and can be made as flexible sheets. The original conducting polymer was polyacetylene made by polymerising acetylene (ethyne) and consists of chains with alternate double and single bonds. As with silicon, it has a full valence band and an empty conduction band separated by a small band gap, making it a semiconductor. Dopants can be added to form n-type and p-type semiconductors. While polyacetylene itself is susceptible to attack by oxygen, there are other conducting polymers which could form the basis of solar cells. Figure 2.5 shows the monomers of some of these polymers. The polymers are cheap to make. They also have the advantage of being flexible. These polymers can also be used in gas sensors and LEDs.

CdTe solar cells are similar in structure to silicon cells (Figure 2.6) but do not require an aluminium frame as they are held together by glass plates top and bottom. The sunlight is mostly absorbed by the thin CdTe layer. Cadmium is obtained as a by-product from mining zinc ores and tellurium from copper ores. Cadmium is toxic and under rising threat

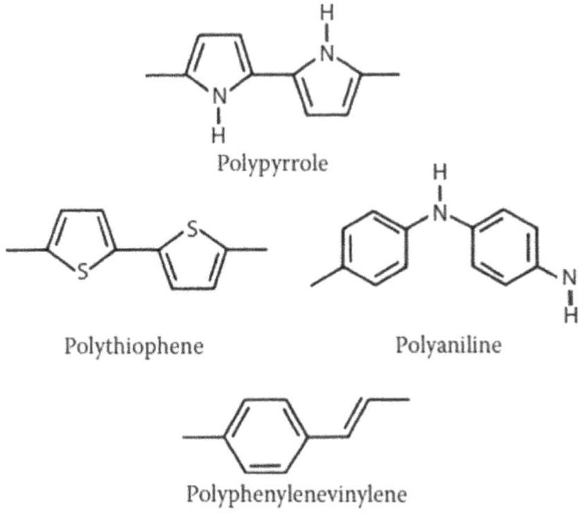

Polypyrrole

Polythiophene   Polyaniline

Polyphenylenevinylene

Figure 2.5 **Examples of conducting polymers. The monomers of these polymers are shown.**

glass
$SnO_2Cd_2SnO_4$
CdS
CdTe
C and metal

Figure 2.6   Cadmium telluride solar cell.

of availability from increased use. Tellurium is under threat of availability in the next 100 years. Both can be recycled.

Thin film gallium arsenide solar cells are the most efficient but the expense of producing the materials has hindered their widespread use. They have, however, been used on spacecraft since 1965, when they were used on the Russian Venera spacecraft. Recent developments include multijunction solar cells and triple junction cells, which were used on the Mars Exploration Rovers. Multijunction solar cells use more than one type of semiconductor. The different materials absorb light at different frequencies and so more of the light from the sun can be used. The triple junction cell uses GaInP, GaInAs and Ge. Gallium, arsenic and indium are under threat of availability in the next 100 years. Arsenic is toxic but gallium is not. Germanium ores are scarce. The element is obtained from zinc ores and is non-toxic.

Dye-sensitised solar cells were proposed in the 1960s and have attracted much research but are not commercially available, mainly because of the liquid electrolyte which could freeze at low temperatures and expand at high temperatures. In addition, the dye is usually a compound of ruthenium and there is a platinum electrode. The availability of both ruthenium and platinum is under threat due to increased use. Platinum is expensive.

Figure 2.7 shows such a cell.

The way these cells work differs from those discussed in Section 2.1.2. A titanium dioxide layer is coated with a dye.

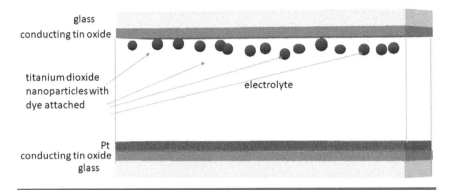

Figure 2.7 **Schematic diagram of a dye-sensitive cell.**

The dye absorbs sunlight and in its excited state transfers an electron to the titanium dioxide and then to a transparent electrode of fluorine doped tin oxide. The electrons return via the external circuit to the counter electrode. This counter electrode is commonly made from platinum and provides electrons to convert triiodide in the electrolyte to iodide. The iodide then reforms the triiodide, giving electrons to the dye.

$$Dye + hf \rightarrow Dye^* \rightarrow Dye^+ + e^-$$

$$I_3^- + 2e^- \rightarrow 3I^-$$

$$Dye^+ + e^- \rightarrow Dye.$$

Hybrid perovskite cells are not yet available commercially. Figure 2.8 shows a proposed solar cell.

Figure 2.8 **One proposed hybrid perovskite cell.**

Sunlight is absorbed by the perovskite. Electrons are donated to the TiO$_2$ layer and replaced by electrons from the hole transport medium. Hole here refers to the absence of an electron. The process is shown in Figure 2.9.

A commonly used hole transport medium is spiro-OMeTAD. Figure 2.10 shows the chemical structure of this compound, an organic semiconductor.

Hybrid perovskites are fragile and subject to attack by air and water, so that although they show promise due to their high efficiency, there are problems with their long-term

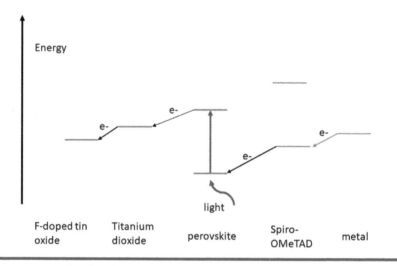

Figure 2.9 Sketch of processes occurring in a perovskite solar cell.

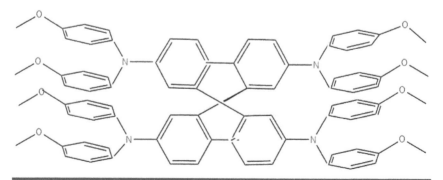

Figure 2.10 Structure of spiro-Ome TAD.

stability. Lead is toxic. The contacts are silver or gold which are expensive. A serious threat to silver supplies in the next 100 years is forecast. Gold comes from conflict minerals and availability is limited.

2.2 Other Considerations

The obvious requirement is sunlight. Solar cells do not produce electricity at night and so you need an alternative source or a way of storing excess energy produced during the day. However, you do not need the levels of sunlight occurring in the tropics. Solar cells are extensively used in the UK on rooftops and in solar cell farms. Efficiency is improved by having the solar cell south-facing in the northern hemisphere and north facing in the southern hemisphere and at an angle that will maximise the sunlight falling on the cells. This will depend on the latitude of the site. They also need to be in a position that is not shaded by trees or nearby buildings.

Current solar cells have a lifetime of about 30 years. Indeed at least one company offers a 15-year guarantee. Maintenance costs are low and at the end of life of the cell you saw in the previous section that recycling of many components is possible. Solar panels made as thin, flexible sheets such as those using conducting organic molecules have a shorter lifetime (closer to 15 years).

To use solar cells as part of the input to a national grid, the power produced by the cell must be converted from direct current (DC) to alternating current (AC) and software employed to regulate the passing of power from the solar cell to the National grid.

A disadvantage is the area required. In central Europe, about 1000 kWhm^{-2} are received from sunlight and in equatorial Africa and parts of Australia about 2200 kWhm^{-2}. On average, a solar cell is 21% efficient. The average medium size household in the UK (three to four bedrooms) uses 2900 kWh of electricity per year. Assuming the central European input of sunlight, solar cells provide 210 kWhm^{-2}, such a household would need about 14 square metres of solar cells to provide their current electricity consumption. In Australia a similar household would only need about 6 square metres.

2.3 Solar Thermal Panels

Solar thermal panels work by using sunlight to heat water which is then fed into the hot water tank or used to heat the water in the tank. This reduces the amount of gas or electricity needed to provide hot water for baths, showers, washing up, etc. Such panels have been used for decades. For example, one UK firm has been installing thermal panels since 1981. There are a number of designs.

In active indirect closed loop systems, the panels contain an array of copper tubes filled with water and an antifreeze on a black metal absorbing sheet. Once the water–antifreeze mix reaches a certain temperature, it is pumped into coils within the hot water cylinder where the heat is transferred to the hot water. The cooled water–antifreeze is then returned to the panel.

In active direct open loop systems, cold water is taken from the hot water tank and fed to the panels where it is heated and returned to the tank. Where there is a danger of the water in the panels freezing, a controller cuts off the supply to the panels.

In active drainback systems, water from the hot water tank is pumped to the panel for heating. When the panels are not being used, the water drains back to the tank to avoid freezing.

2.3.1 Materials

Copper is used for the tubes in which the water or water–antifreeze is heated and for tubes at the top and bottom of the panel. The antifreeze used is usually glycol. Insulation is needed between the tubes and the roof. The absorbance sheet is dark and can be a polymer, copper, aluminium or steel. The panel has a transparent cover, usually glass.

2.3.2 Other Considerations

As with photovoltaic solar cells, solar thermal panels need to be installed on sun-facing roofs at an angle to maximise the amount of sunlight reaching them. In a typical UK summer, they can supply all the hot water needed for a small family.

They are expensive to install but maintenance is cheap and they can last 20–30 years.

They do require, however, a hot water tank, so cannot be used with combi boilers, and need a backup water heating system such as an immersion heater, boiler or heat exchanger for times when they produce insufficient hot water, for example, during winter. It is best to use the hot water to shower, wash up, etc. in the evening as this gives the system the maximum time to work.

Questions

1. Cadmium telluride (CdTe) solar cells often contain Cu atoms as a dopant. Would this produce a p-type or n-type semiconductor?
2. The probability of an electron in a silicon solar cell promoted from the valence band to the conduction band emitting light and returning immediately to the conduction band is very low. Why is this important?
3. Which of the photovoltaic cells described here have a different mechanism to that in Figure 2.3?
4. Which of the photovoltaic cells have a thin layer of the solid responsible for the photovoltaic effect?

5. Mining and transporting minerals use energy and currently involve emission of carbon dioxide. What minerals are needed for hybrid perovskite solar cells?
6. Which elements needed for dye-sensitive solar cells are in plentiful supply? See bit.ly/euchems-pt for availabilities.
7. Energy and availability of minerals have been mentioned. What other resources need to be taken into account?

Chapter 3

Wind Power

This chapter looks at how the power of the wind can be used to provide energy. Wind power has been used for centuries in windmills. Most windmills were used to grind grain. In windmills, the wind blowing against the sails causes them to rotate. This vertical rotation is transferred via gears to a long pole, which rotates horizontally and is used to grind the grain. In the Netherlands, windmills were also used to pump water to prevent flooding of the dykes and for industrial processes. There are still windmills in use, but here we will look at wind turbines which are used to produce electricity.

3.1 Wind Turbines

Wind turbines work on a similar principle, but the rotation of the sails or blades is used to generate electricity. Most wind turbines have three blades. The set of blades faces the wind and the blades are designed in a similar manner to aircraft wings. Figure 3.1 shows a modern blade.

Aircraft wings have a curved upper surface. This leads to the air above the wing travelling faster than that below the wing, and as a result the pressure is less on top. This produces

DOI: 10.1201/9781003294337-3

Figure 3.1   **This photo shows one of the three 135-ft (41 m) blades of a turbine before installation (Paul Cryan U.S. Geological Survey).**

a lifting force at right angles to the wind direction and dependent on the wind speed. There is also a force at right angles to the lifting force called drag. The wing needs to be designed so that the lift exceeds the drag. Note from Figure 3.1 that wind turbine blades also have a curved upper surface. In addition, they are tapered towards the tip.

With wind turbine blades the lift depends on the effective wind speed, which combines the actual wind speed with the movement of the air due to the rotation of the blades. Figure 3.2 illustrates this effect.

This effective wind speed varies along the length of the blade as the speed of the blade with respect to the air varies at different positions along the length. Blades are built in sections, with the surface of the blade varying along the length to maintain an optimum effective angle for each section. They are also tapered so that the diameter at the far end is small. This is to ensure that the air is slowed by the same amount along the length of the blade.

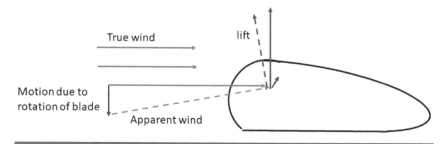

Figure 3.2   The effect of rotation on the lift of a wind turbine blade. The dashed lines show the apparent wind and the direction of lift due to this.

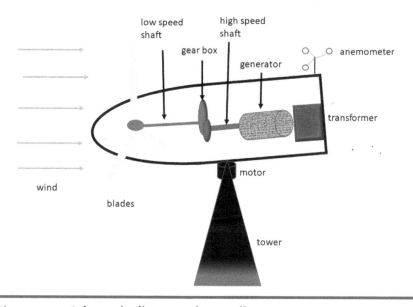

Figure 3.3   **Schematic diagram of a nacelle.**

The structure on top of a wind turbine to which the blades are attached is known as the nacelle. Figure 3.3 shows a schematic diagram of a nacelle. On top of the nacelle is an instrument, an anemometer, to measure wind speed and direction. This connects to a motor that alters the direction that the blades are facing as the wind varies to obtain maximum efficiency. It is also linked to a brake which stops the blades from rotating when the wind speed is too high to prevent damage to the turbine (brakes are also used in windmills for this purpose). As the blades rotate, they turn a spindle. The spindle rotates quite slowly so gears are used to link it to a rod which turns around about 80 times faster. The rod in turn is linked to an electrical generator.

Wind turbines to produce power for National Grids are mounted on towers 40–100 m high. This enables them to use winds that can be stronger than those at ground level and free from obstruction by trees, houses, etc. These towers are either hollow towers or lattice structures made of steel and on a concrete base. Wires carry the current produced by the generator from the nacelle to the base of the tower. Blades are about 15–60 m long and hollow. They are often made from fibre glass, carbon fibre or composites as they need to be light and strong. Currently (2021) the largest wind turbine is 260 m high and had blades 107 m long. This turbine has the capacity to produce 12–14 MW of power.

There are also smaller turbines for use by individual homes or on roadsides. They work on the same principles but can be used to produce direct current (DC) for charging batteries, for example, rather than the alternating current (AC) needed for the National Grid. These can be small versions of the wind turbines in wind farms. However, there are some that have a different type of blade which rotates around a vertical axis and can catch the wind from any direction. Figure 3.4 shows an example of such a turbine.

3.2 Electrical Generators

Generators work on the principle that a moving magnetic field will induce an electrical current in a conductor. They are used not only in wind turbines but also for the generation of power from waves and tides, hydroelectric dams and power stations using fossil fuels and biomass. In producing power for the

Figure 3.4 **A small Giromill Darrieus vertical-axis wind turbine (VAWT) in rural Rollinsville, Colorado (Tony Webster). https://creativecommons.org/licenses/by/2.0/legalcode.**

National Grid, the conductors are usually coils of copper wire surrounding a rotating array of permanent or electro-magnets. As the magnets rotate, the copper coil is affected alternately by the north and south poles of the magnets. This causes the electrical current to alternate its direction of flow in the manner of a sine wave, as shown in Figure 3.5, giving an AC current.

The speed at which the blades rotate in wind turbines varies with the strength of the wind, and hence the frequency of the AC current varies. This current therefore must be transformed to give an AC current of constant frequency that corresponds to the frequency of the current needed for the National Grid. This is done by first transforming the original AC current to a DC current and then to an AC current with the grid frequency.

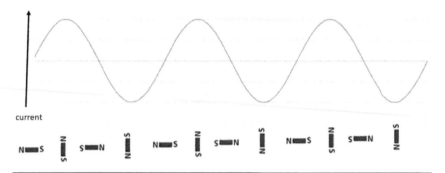

current

Figure 3.5 **Variation in the electrical current with time as the magnets rotate. The rotation of the magnets is indicated by bar magnets.**

3.3 Materials

The tower of wind turbines is usually made of tubular steel because it has to be robust; off-shore wind turbines, for example, are not only exposed to high winds but are also buffeted by the sea. For further protection against corrosion, the towers are painted or coated with zinc. The rotor, gears and generator are also mainly made of steel or iron. Iron is a readily available element. Steel is an alloy of iron with percentages of carbon and other metals depending on its use. Stainless steel as used in the towers is based on austenite. This is a face-centred cubic form of iron. Figure 3.6 shows its unit cell. It is an energy-intensive material to produce, and carbon dioxide is emitted in the process. Research is looking at ways of making steel that do not emit carbon dioxide and of capturing the carbon dioxide produced. Steel is recyclable.

Concrete is used to form a base for the tower. Traditional concrete is made by binding aggregates such as sand, gravel or crushed stone with a cement-water paste. Cement is currently responsible for a significant emission of carbon dioxide (approximately 8% of global emissions). The widely used Portland cement can be made by heating limestone (calcium carbonate) with clay (layered aluminosilicates} to 1450–1500°C, resulting in the formation of calcium silicate,

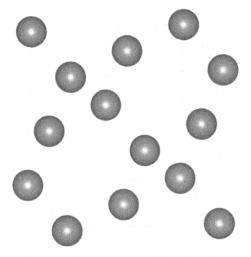

Figure 3.6 Atom positions in a face-centred cubic crystal.

calcium aluminate and carbon dioxide. The resulting mixture is allowed to cool and is then ground and mixed with gypsum (calcium sulfate dihydrate).

Figure 3.7 shows an idealised clay structure.

Concrete can last for 50–100 years, and over this period it absorbs 10–30% of the carbon dioxide emitted in its manufacture. Replacement materials, such as low-grade clay in the cement, using less cement in the concrete or changes in production methods such as using less fossil fuel in the kilns can reduce the carbon dioxide emissions. Using less concrete and recycling concrete also help.

The blades need to be strong but light and are often made of fibre glass. This is hard to recycle, although methods of recycling are being developed. For example, a German company is using discarded blades to substitute for sand in concrete manufacture. In some cases, it is possible to re-use blades from decommissioned wind turbines in new turbines. Fibre glass is also used for the nacelle shell.

Then of course there are the metals in the generator, and copper wire is used in the windings and in cables linking the nacelle to the ground and to other turbines. Permanent

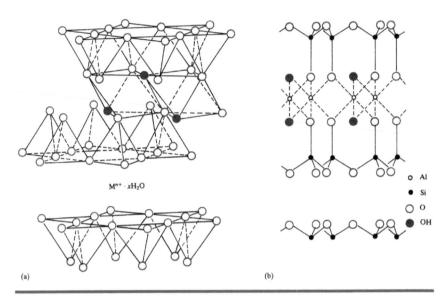

$M^{n+} \cdot xH_2O$

o Al
● Si
○ O
● OH

(a)   (b)

Figure 3.7   **Idealised structure of the layers in a clay. (a) Showing only the oxygen/hydroxyl framework and (b) also showing the aluminium and silicon positions.**

magnets can be $SmCo_5$ or NdFe, which contain rare earth elements and, in the case of $SmCo_5$, cobalt, whose main source is mines in the Democratic Republic of the Congo. Iron is an abundant metal. Supplies of copper have become limited due to the widespread use and rising price of the metal.

3.4 Other Considerations

Wind turbines can only be efficiently used when the wind is neither too light nor too strong. A steady wind is ideal.

The turbines have a large 'footprint' at ground level. In a wind farm the turbines should be sited at a distance of five times the diameter of the circle formed by the rotation of the blades from adjacent turbines. This ensures that any turbulence produced at one turbine does not affect its neighbours.

Wind turbines can be noisy and so are not generally built near where people are living. Thus large wind turbine farms are built in remote areas or off-shore. There have been concerns about the effect of these turbines on birds. Overall far more birds are killed by pet cats than by wind turbines. However, some remote sites may be areas of special interest which are breeding grounds for some rare species of birds. Manufacturers are working on reducing the noise and making the blades less dangerous for birds.

At high wind speeds, the blades can be eroded by rain.

On average, a wind turbine lasts 20–25 years. While the concrete bases could last 50–100 years, they cannot be re-used for replacement wind turbine towers.

Questions

1. A wind turbine has blades 30 m in length. The rotor is rotating at 15 revolutions per minute (rpm). At what speed in ms^{-1} is the tip of the blade moving?
2. Why does the frequency of the alternating current produced by the generator vary?
3. Stainless steel is strong and resistant to corrosion. What other property does it have that makes it a useful material for items in the nacelle?
4. A wind turbine has blades 30 m in length. How far apart should a farm of such turbines be spaced?
5. In 2020, the total capacity of wind turbines in the UK was about 25,000 MW. The electricity generated was about 75,000 GWh. If the turbines produced 25,000 MW continuously, then they would have produced around 220,000 GWh. Suggest some reasons for the lower value generated.

Chapter 4

Water Power

This chapter covers ways of using the motion of water to provide energy. Just as wind power has been used for centuries to grind corn in windmills, so water power has been used in water mills. Water power has been used to generate electricity using hydroelectric dams for over a century. Tides can also be used to generate power. A tidal power plant has been running since the 1960s in France, although the use of tidal power plants otherwise has only been investigated recently. More recently generators using waves are starting to be introduced.

Hydroelectric power provides the greatest share of renewable energy, mostly via hydroelectric dams. Norway produces around 90% of its electricity using this method.

4.1 Hydroelectric Dams

Hydroelectric dams have been used to generate electricity for over a century. Figure 4.1 shows a photograph of the Cheoah dam in the USA, which was completed in 1919.

To the right of the photo, note the yellow building. This is the power house, and from here electricity is fed to the pylons shown and then onto to National Grid.

DOI: 10.1201/9781003294337-4

Figure 4.1   **Cheoah hydroelectric development. (Dantripphoto 1 August 2010. This file is licensed under the Creative Commons Attribution-Share Alike 3.0 Unported license.)**

The largest hydroelectric dam in the world is the Three Gorges Dam on the Yangtse river in China. This is 185 m high and can generate 22,500 MW of power.

Large hydroelectric dams are built on rivers which have sections where the river drops steeply. A dam is built at the high point of the drop, and a reservoir of water forms behind it. A channel within the dam, called a penstock, takes water from the reservoir to the river below. The amount of water going from the reservoir to the penstock is controlled by sluice gates which can be raised and lowered to regulate the flow of water from the reservoir. Towards the bottom of the penstock, there is a turbine attached to a shaft. The shaft is connected to a rotor-forming part of an electrical generator (see Chapter 3, Section 3.1). A direct (DC) electric current produces a magnetic field in the rotor and this induces an alternating (AC) electric current in coils of conducting wire attached to a stationary ring known as the stator. The coils are attached to wires which carry the current to a transformer and thence to the

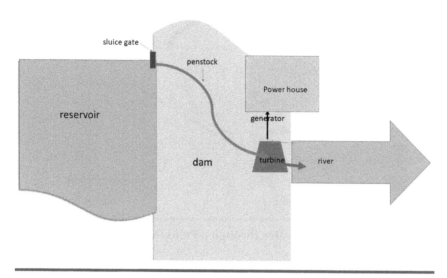

Figure 4.2   Schematic diagram of a hydroelectric dam.

National Grid. Figure 4.2 shows a schematic plan of a typical hydroelectric dam.

There are two types of water turbines in use. Impulse turbines are driven by the kinetic energy of water exiting from the penstock through a nozzle. Reaction turbines are driven by both the pressure of the water entering the turbine and the kinetic energy. These have to be completely enclosed and submerged. The pressure depends on the distance the input water has to fall to reach the turbine.

Hydroelectric power plants using high dams usually use either a Pelton wheel or a Francis turbine.

The **Pelton wheel** is an impulse generator developed in the 19th century. It consists of a flat plate attached to a horizontal rotating shaft at right angles to the surface of the plate. Attached to the plate are a number of cup-shaped blades known as buckets. Water from the penstock is forced through a nozzle and hits the buckets, causing the wheel to rotate. Figure 4.3 shows the sketch of the arrangement.

Pelton wheels are suitable for dams with a large drop (400–2000 m).

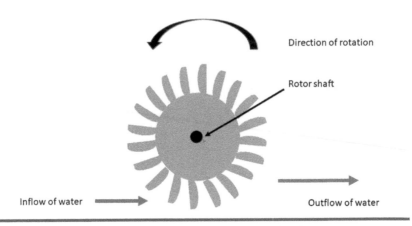

Figure 4.3   **Passage of water through a Pelton wheel.**

The **Francis** turbine is a reaction turbine. This has blades arranged around a vertical rotating shaft. A spiral shell surrounds the turbine taking water from the penstock to the turbine. This reduces in diameter from the penstock to the turbine to maintain the pressure. Guide vanes take the water to the turbine blades. These control the angle at which the water strikes the blades and the rate of flow of the water entering the turbine. Water exits the turbine downwards (Figure 4.4).

Francis turbines are used for hydroelectric plants with a drop of 3–600 m.

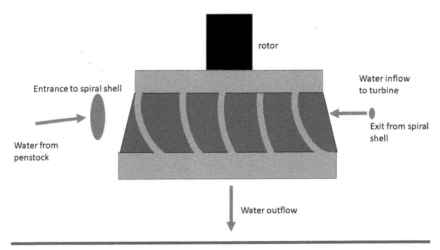

Figure 4.4   **Passage of water through a Francis turbine.**

4.2 River Turbines

It is possible to generate electricity from a turbine placed directly in a channel coming from a river. The water flows from the channel into a pipe and over a turbine, causing it to rotate. Kaplan turbines are generally used. These are similar to ship's propellers. As for hydroelectric dams, the turbine is linked to a generator.

4.3 Wave Power

Waves are made by winds passing over the oceans. Because the kinetic energy of waves is far greater than that of wind, wave power generators can be much smaller than wind turbines. The world's first commercial wave farm was set up on the coast of Portugal in 2008 and delivered 2.25 MW of electricity. Technical and economic problems caused this facility to be shut down. Work is continuing to develop wave power generators, however, as they could be potentially cheaper than wind turbines. In 2021, a wave power generator in the Orkney Islands, Scotland, started providing power to the National Grid. Wave power generators use the up and down motion of waves to produce electricity. The vertical wave motion is used to drive a piston to compress a fluid in a cylinder. The fluid, which can be oil or air, is then fed into a hydraulic motor where the linear motion is converted to rotation about a shaft which is attached to a generator. When the fluid is air it drives a wind turbine. However, as the piston goes up and down, the direction of flow is reversed and the turbine must be able to work in either direction.

4.4 Tidal Power

Tidal power was used by the Romans to grind corn, and the first use to generate electricity was in France in the 1960s.

Tides involve the large-scale movement of water in the oceans and seas. They are caused by the gravitational pull of the moon. This causes a bulge in the water on the side facing the moon due to increased gravitational attraction and on the opposite side due to the decrease in the pull. As the Earth rotates, different places experience the effect of the moon's gravity. Since the bulges occur both on the side facing the moon and the side opposite, this leads to two high tides per day in most places.

So, how is this movement of the water harnessed to generate electricity? There are three methods in use.

4.4.1 Tidal Barrages

Tidal barrages are large straight structures placed in estuaries to form a reservoir at high tide. Below the main structure there are tunnels which capture the water as it flows in and out and passes it over a turbine. The amount of electricity produced depends on the height difference between high and low tides. For a sizeable output, this difference in height should be at least 3 m.

4.4.2 Tidal Lagoons

Tidal lagoons are similar but follow the coast line, and turbines are placed so that the water causes them to rotate as the lagoon fills and empties. This type was used in the first tidal power plant on the River Rance in Brittany, France.

4.4.3 Tidal Streams

Tidal streams are fast-flowing currents whose energy can be captured by turbines placed on the ocean floor. These turbines are much larger than wind turbines and use propellers.

4.5 Materials

Hydroelectric dams are generally made of rock and sediment and other locally available materials with a concrete coating. In Chapter 3, we saw that concrete manufacture was a large source of carbon dioxide emissions. Turbines are made of steel with a high percentage of chromium. Steel was discussed in Chapter 3. Chromium is under threat of future shortages due to increased use.

4.6 Other Considerations

Building the reservoir for a hydroelectric dam leads to flooding of a large area behind the dam. This can lead to the destruction of wildlife habitats. Those of tigers, for example, are believed to have been affected. Fish such as salmon that swim upstream to spawn cannot leap directly up the dam. However, fish ladders can be added to the side of the dam. The reservoirs can also lead to a build-up of algae and the emission of methane from rotting vegetation.

Displacement of people from their homes can occur. For example, reservoirs can drown villages, causing the inhabitants to be re-housed. It is estimated that the creation of the Three Gorges Dam involved the relocation of 1.4 million people. The high Aswan Dam on the Nile in Egypt caused 90,000 Egyptian and Sudanese people to be relocated. The building of this dam also became famous as the reservoir would have submerged the ancient temple of Abu Simbel. This temple and its complex were moved at great expense.

The dams reduce the flow of water in the river below. The Hoover Dam in the USA caused the Colorado river to end before it reached the sea. The recent dam built on the Nile in Ethiopia has worried the countries downstream, Egypt and Sudan, which depend on water from the Nile for agriculture.

On the plus side the dams can regulate the flow of water for agriculture and prevent flooding.

Waves are always present but not everywhere. The best sites for wave power are coasts on the western extremities of continents. The structures are situated close to the shore. This may disrupt fish and also shipping.

The timing of tides is predictable, but they are not continuous and tidal generators only produce electricity for about ten hours in a day. The height of the tide also varies. To capture the flow of water, turbines need to be reversible to catch the tides incoming and outgoing.

Putting barriers across estuaries to form tidal barrages leads to environmental problems such as reduced diversity of birds and fish and areas of tidal flats that are too salty or mainly fresh water.

Tidal streams perform best in shallow water. The turbines on the sea floor turn slowly, and this helps fish evade them. This method is thus more environmentally friendly.

Questions

1. Solar cells convert light to electricity. What forms of energy are used to produce electricity from hydroelectric dams?
2. Which type of water power generators can produce electricity continuously and which produce intermittent power?
3. How are (a) tides and (b) waves produced?
4. Describe some of the environmental impacts of (a) land-based water power systems and (b) coastal-based systems.

Chapter 5

Geothermal Energy

This chapter is concerned with energy from heat within the Earth. It covers spectacular features such as geysers and hot springs but also the use of heat in layers only a few metres beneath our feet. It looks at how heat from all these sources can be used for heating buildings and for producing electricity.

5.1 The Origin of Geothermal Energy

Figure 5.1 shows the main layers of the Earth's interior.

Although the Sun warms the surface, this heat does not spread down very far. The inner layers are heated by friction and gravitational forces from the formation of the Earth and by the decay of radioactive elements in the mantle. The core is mainly iron and nickel – solid in the inner core and molten in the outer core. The temperature increases from the crust through the mantle to 3000°C and then increase more rapidly from the deepest mantle to the outer core, reaching 3800°C. It then increases more slowly through the core, finally reaching 6000°C.

The main radioactive elements are thorium, uranium and potassium. Radioactive isotopes of these elements decay to

DOI: 10.1201/9781003294337-5

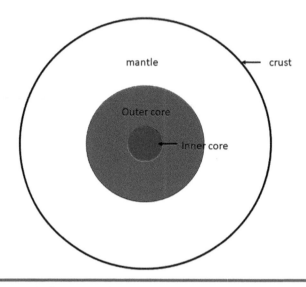

Figure 5.1   Main layers of the Earth's interior. Not to scale.

form other elements by randomly emitting particles such as helium nuclei (alpha particles), electrons or positrons (beta particles) and/or gamma rays and neutrinos. The random decay results in an exponential decrease (Figure 5.2) in the concentration, which can be characterised by the time taken to reduce the concentration to half its initial value. This time is called the half-life.

The most stable isotope of uranium, ^{238}U, has a half-life of 4.5×10^9 years. The most stable isotope of thorium, ^{232}Th, has a half-life of 14.06×10^9 years.

Figure 5.3 shows the decay path of these isotopes. The isotope of potassium with mass number 40, ^{40}K, has a half-life of 1.25×10^9 years. It decays to ^{40}Ca and ^{40}Ar. The decay to ^{40}Ar is used to date rocks.

How does radioactive decay lead to heating the core? In alpha or beta decay, the radioactive isotope spontaneously emits an alpha particle (helium nucleus) or a beta particle (an electron). You might expect the masses of the new nuclide plus the alpha particle or beta particle to add up to the mass of the original isotope. However, this is not true. The sum of

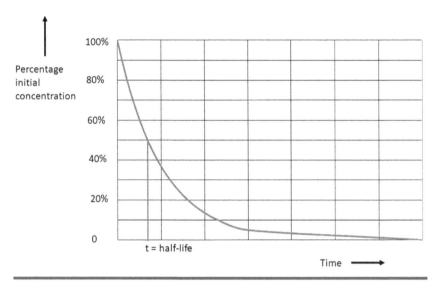

Figure 5.2   The decay of the concentration of a radioactive nucleus as a function of time.

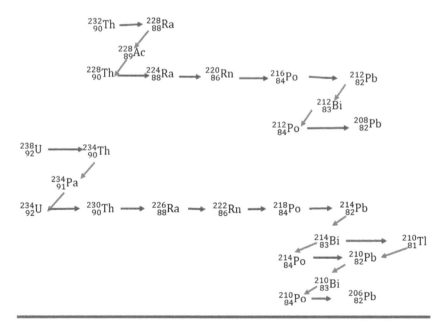

Figure 5.3   Radioactive decay paths for ^{232}Th and ^{238}U. Red arrows indicate alpha emission and blue arrows indicate beta emission.

the product masses is less than the mass of the nucleus of the original isotope. The missing mass has been converted to energy, mainly kinetic energy of the emitted alpha and beta particles. The amount of mass converted to energy can be calculated using Einstein's famous equation $E = mc^2$.

The core is very dense, so the particles do not escape to outer layers, but the kinetic energy is converted to vibrational energy of the core materials resulting in a rise in temperature. The heat is then transferred to the mantle.

5.2 Accessing Geothermal Energy

We cannot access the core and must rely on the heat in the crust and upper mantle. A few metres below the surface of the Earth, there are pockets with temperatures of around 150°C. This is one source. At higher temperatures of around 1000°C, rocks in the mantle can partially melt to form magma. Magma can reach the surface via volcanoes but can also heat underground aquifers. The hot water can reach the surface, for example, as geysers and hot springs. Another source is the water in disused mines.

5.2.1 Using Geothermal Energy for Heating

The low-temperature geothermal energy found just metres below the surface or in hot springs has been used for centuries for heating houses, greenhouses, industrial processes and cooking.

Iceland uses both low-temperature and high-temperature sources. The country lies across the boundary of the North American and Eurasian tectonic plates, the mid-Atlantic ridge. The plates are moving slowly apart, and as they do so, mantle material wells up into the ridge, making Iceland one of the most active places on Earth with many volcanoes and hot springs. In the active area, there are at least 20 areas containing steam fields with temperatures reaching 250°C within 1 km below ground. Away from the volcanic activity there are over 600 hot springs with temperatures greater than 20°C. Around 90% of Iceland's houses are heated with geo-thermal energy.

Low-temperature resources can be used almost anywhere. To use these resources to heat a house, tubes containing water or water plus antifreeze are installed under the garden. When accessing an underground water source, these tubes are open-ended. Closed loops are used to access solid heat sources. Inside the house, these tubes transfer heat to another set of tubing containing refrigerant. The refrigerant flows through a compressor. Compressing the fluid leads to a rise in its tem-perature. The heated fluid then transfers heat to water pipes which can provide underfloor heating, heat radiators or a hot water tank. The pump and condenser are powered by electricity.

Larger buildings than houses can also use geother-mal energy in this way. For example, the Oxford Brookes University in Oxford, UK, is currently constructing a 500 kW geothermal plant to provide heating and cooling for its campus.

5.2.2 Using Geothermal Energy to Produce Electricity

To produce electricity the geothermal heat is used to produce steam. The steam then drives a turbine which is attached to a generator in the same way as fossil fuel and biomass power plants. After passing through the turbine, the steam is scrubbed to remove sulfur compounds such as hydro-gen sulfide (H_2S) and then condensed, cooled and returned underground.

In volcanically active countries such as Iceland and the Philippines, flash steam power plants can be used. Underground sources of water with temperatures above 182°C are pumped to a lower pressure area where the water turns to steam. The steam is then fed to a turbine. The Hellisheidi power plant in Iceland produces about 303 MW electricity and 400 MW heat in a combined heat and power plant.

It is also possible to use underground sources where the water is present as steam. In California a complex of 13 power plants uses steam reservoirs to generate about 725 MW of electricity.

Geothermal power plants can also operate in less volcanically active regions. The Eden project in Cornwall, UK, is currently constructing a geothermal combined heat and power plant to heat its Biome, greenhouses and offices. Cornwall is a particularly favourable part of the UK for geothermal projects because the underlying rock is granite. Granite is an igneous rock formed by liquid magma as it cools. It is a mixture of silicate minerals, mainly quartz (SiO_2), feldspar ($M(Al.Si)_4O_8$) and mica ($KM_3(AlSi_3O_{10}(OH)_2)$). The heat flow in Cornwall granite is particularly high, and it has been estimated that temperatures in excess of 200°C can be reached just 5 km below the surface. So what is responsible for the heat in granite? Granite contains small amounts of radioactive uranium (2–50 ppm) and thorium (8–56 ppm) ions. Granite also contains 2.5–4.5% of potassium, but only 0.01% of this is the radioactive isotope ^{40}K. As you saw in Section 5.1, these isotopes decay, releasing energy, and it is this energy that warms the granite.

5.3 Considerations

Geothermal power plants can operate continuously; they do not depend on the weather.

While most places on the Earth can be used for low-temperature heat source plants, places with high-temperature sources are less common.

The area of land required for a given size power plant is much less than for wind, photovoltaic or coal plants.

Electricity is needed to run the pumps and condensers, and this may be supplied by sources which emit carbon dioxide.

Drilling into the ground to test for possible sites and to prepare for the tubes to access the heat source requires heavy machinery which may use fossil fuels for their operation.

Some geothermal plants have been linked to subsidence.

Questions

1. ^{232}Th has a half-life of 14.06×10^9 years. How many years would it take to reduce the concentration of this isotope to a quarter of its initial value?
2. The outer core is thought to be responsible for the Earth's magnetic field. Thinking about how generators work, how could the outer core produce a magnetic field?
3. What is the effect of increasing the pressure on a fluid confined in a container?
4. Why is the Philippines a good place for geothermal energy plants?
5. A radioactive gas can accumulate under houses built on granite. From the decay series of Th and U, which gas is this likely to be?

Chapter 6

Hydrogen

Replacing fossil fuels with hydrogen is an attractive idea. Burning hydrogen produces only water.

$$2H_2(g) + O_2(g) = 2H_2O(l) \qquad (6.1)$$

Pilot projects adding a percentage of hydrogen to the natural gas (methane) delivered to homes have taken place, and hydrogen-powered vehicles have been around for some years. However, there are no deposits of hydrogen gas and so the hydrogen has to be manufactured. In this chapter, we discuss the ways of preparing, storing, transporting and using hydrogen gas.

6.1 Hydrogen Production

Production methods are categorised by colour according to how much carbon dioxide is emitted in the process.

One way to obtain hydrogen gas is to electrolyse water, that is, passing an electric current through water to reverse Equation 6.1. This method can only be classed as clean or renewable if the electricity is provided by a clean or renewable

DOI: 10.1201/9781003294337-6   **45** source. Hydrogen produced by electricity from a renewable source is known as green hydrogen. If the electricity is provided by nuclear power, it is classified as pink or red hydrogen. Hydrogen produced using natural gas to generate electricity for the electrolysis is classified as grey hydrogen. Hydrogen produced using coal-fired power is known as brown hydrogen.

Currently, much hydrogen is prepared through the chemical reactions of carbon-containing molecules. This produces greenhouse gases as a by-product. When combined with carbon capture, it is known as blue hydrogen. Carbon capture is discussed in Chapter 9.

A relatively new method still under development is methane pyrolysis. The products are carbon and hydrogen rather than carbon dioxide and hydrogen. The hydrogen is less pure than that produced by electrolysis and is known as turquoise carbon if renewable energy sources are used to effect the pyrolysis.

Other methods include photolysis and enzyme reactions.

6.1.1 Electrolysis

There are three main types of electrolyser – alkaline, PEM and solid oxide.

In **alkaline hydrolysis**, an electric current is passed through an aqueous solution of potassium hydroxide. Hydroxide ions travel from the cathode to the anode, and hydrogen is produced at the cathode.

$$H_2O(l) = H^+(aq) + OH^-(aq) \quad \text{dissociation of water in solution}$$

$$OH^-(aq) + H^+ = H_2O(l) \quad \text{anode reaction}$$

$$2H^+(aq) + 2e = H_2(g) \quad \text{cathode reaction}$$

where e stands for an electron from the external circuit.

A membrane between the electrodes separates hydrogen and oxygen. Figure 6.1 is a schematic diagram of an electrolytic cell.

PEM stands for polymer electrolyte membrane.

Water reacts at the anode to form hydrogen ions and oxygen. Between the two electrodes is a thin proton exchange membrane. This allows hydrogen ions to pass through selectively. At the cathode, the hydrogen ions react with electrons from the external circuit to form hydrogen gas.

Figure 6.2 is a schematic diagram of a PEM electrolysis cell.

$$2H_2O(l) = O_2(g) + 4H^+(aq) + 4e \quad \text{anode reaction}$$

$$2H^+(aq) + 2e = H_2(g) \quad \text{cathode reaction}$$

The polymers that form the membrane have a hydrophobic backbone with hydrophilic groups attached. The hydrophilic groups are responsible for the passage of hydrogen ions through the membrane. These groups are commonly

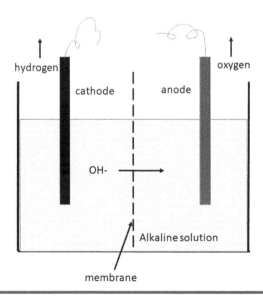

Figure 6.1 **Schematic diagram of an alkaline hydrolysis cell.**

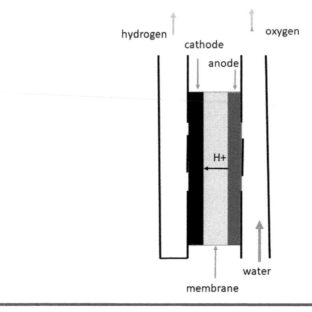

Figure 6.2 **Schematic diagram of a PEM electrolysis cell.**

perfluorosulphonic acid. The chemical structure of a typical membrane is shown in Figure 6.3.

Unlike alkaline hydrolysis cells, these cells can be operated under pressure using a stainless steel container. They can also cope with intermittent power supplies.

Solid Oxide cells contain a solid oxide electrolyte which conducts oxide ions. Steam is used as the source of water and forms hydrogen gas and oxide ions by reacting with electrons at the cathode. At the anode, the oxide ions donate electrons to the circuit and form oxygen gas.

$$2O^{2-}(s) = O_2(g) + 4e \quad \text{anode reaction}$$

$$H_2O(g) + 2e = H_2(g) + O^{2-}(s) \quad \text{cathode reaction.}$$

Figure 6.4 shows such a cell schematically.

These cells operate at high temperatures, typically around 800°C. A typical solid oxide electrolyte is yttrium-stabilised

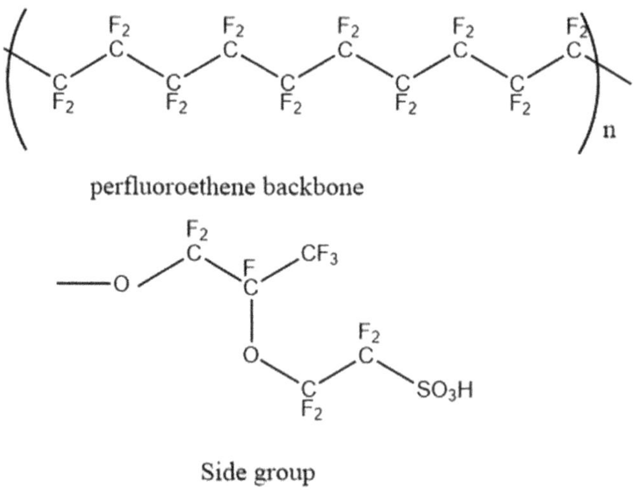

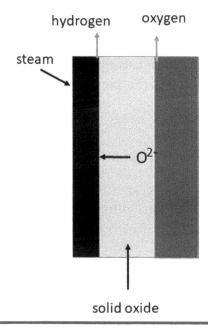

perfluoroethene backbone

Side group

Figure 6.3   Backbone and side group of typical polymer used as PEM membrane. The side group is attached to the backbone in clusters.

hydrogen   oxygen

steam

O^{2-}

solid oxide

Figure 6.4   Schematic diagram of a solid oxide electrolysis cell.

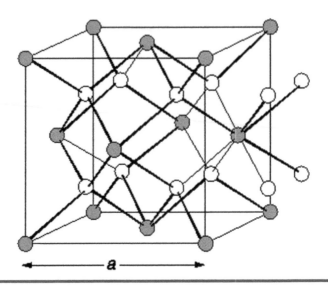

Figure 6.5 Structure of cubic zirconia. Blue spheres zirconium. White spheres oxygen.

zirconia. Figure 6.5 shows the cubic form of zirconia ZrO_2 which is stabilised by replacing some zirconium atoms with yttrium. Because the yttrium ion has a lower charge (Y^{3+}) than zirconium ions (Zr^{4+}), this substitution is balanced by removing oxygen ions. This makes the lattice more open and improves the conduction of oxide ions through the solid.

6.1.2 *Chemical Reactions*

The most widely used reaction is steam reforming in which methane reacts with steam to form hydrogen, carbon monoxide and a small amount of carbon dioxide.

$$CH_4(g) + H_2O(g) = H_2(g) + CO(g)$$

This is currently (2021) used for 50% of hydrogen production globally. Replacing air with pure oxygen produces a clean stream of carbon dioxide. This makes it easier to capture the carbon dioxide and produce blue hydrogen.

Another major source is the water gas shift reaction. In this reaction, water reacts with carbon monoxide to form hydrogen gas and carbon dioxide.

$$H_2O(g) + CO(g) = H_2(g) + CO_2(g)$$

The carbon monoxide from the steam reforming reaction can be used in this reaction.

6.1.3 Biohydrogen Production

Some anaerobic bacteria such as *Clostridium* can produce hydrogen from carbohydrates such as glucose and also from waste biomaterials such as straw and sewage sludge. This process is known as dark fermentation. The method has the attraction of producing hydrogen whilst also using up waste products, but yields are currently low.

An alternative biological method under research is to combine part of the photosynthesis cycle with modified algae that can produce hydrogen.

6.2 Storage and Transport

Storage of hydrogen is one way of storing energy for use with intermittent power supplies.

Hydrogen is a gas at room temperature and pressure with a boiling temperature of –253°C. To store the gas, it must be compressed under high pressure, 700 bar, or cooled to –253°C and stored in an insulated cylinder at lower pressures, 6–350 bar (Atmospheric pressure is 1.01 bar). Currently hydrogen-powered vehicles use cylinders of hydrogen which can be refilled. These cylinders use a carbon fibre composite with a plastic liner and contain hydrogen gas at 700 bars pressure.

An alternative way of storing hydrogen is in solids that will adsorb it or on which hydrogen will adsorb. Interstitial metal hydrides where hydrogen atoms occupy vacant positions in the metal crystal have been suggested as one way of storing hydrogen. Another is the formation of complex hydrides such as sodium aluminium hydride $NaAlH_4$. Porous solids contain channels and cavities that can accommodate molecules and so have been the subject of research for hydrogen adsorption. One such type of solid is a metal organic framework (MOF). MOFs consist of metal ions or clusters of metal atoms joined together by organic ligands. An example is shown in Chapter 9. Another type of solid is activated carbon. The US Department of Energy set a target for room temperature adsorption of hydrogen of 4.5 wt% by 2020.

Hydrogen gas can be transported through pipelines like natural gas. In Germany, a steel pipeline has been used to carry hydrogen at pressures of 20–210 bar since 1938. Hydrogen can cause the steel to become brittle as H atoms are absorbed into the metal. This is less likely for steels based on the face-centred cubic form of iron (Chapter 3, Section 3.2). Hydrogen is more likely than methane to diffuse out of pipelines, but the loss is very small of the order of 0.001%.

Most hydrogen, however, has been transported so far by road or rail in steel cylinders or refrigerated trucks or trailers. Figure 6.6 shows a tank trailer.

Figure 6.6 Hydrogen tank trailer.

6.3 Hydrogen Use

One way to use hydrogen is to burn it directly as a fuel.

However, in current hydrogen-powered vehicles fuel cells are used to produce electricity to drive the motor. The first working fuel cell was made in 1839, and fuel cells have been used on space missions. The first buses powered by fuel cells went into service in 1993, but the cost and the difficulty in storing and transporting hydrogen have meant hydrogen-powered vehicles have not yet become as widespread as those propelled by batteries. Figure 6.7 shows hydrogen-powered vehicles at a filling station.

6.3.1 Fuel Cells

Figure 6.8 is a schematic diagram of a hydrogen fuel cell.

The fuel, in this case hydrogen, and a supply of air are fed into the cell. At the anode hydrogen is oxidised to hydrogen ions and electrons. The electrons travel around the external circuit, and the hydrogen ions pass through the electrolyte to the cathode. At the cathode, the hydrogen ions react with oxygen from the air to give water and heat.

$$H_2(g) = 2H^+(aq) + 2e \quad \text{cathode reaction}$$

$$4H^+(aq) + O_2(g) + 4e = 2H_2O(l) \quad \text{anode reaction.}$$

Figure 6.7 **Hydrogen-powered vehicles.**

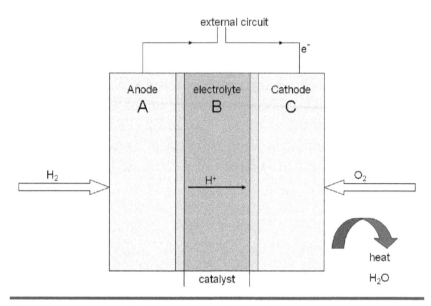

Figure 6.8   **Schematic diagram of a fuel cell.**

As for electrolysis, there are several different types of electrolytes for fuel cells. Some of these – alkali, polymer electrolyte membrane (PEM) and solid oxide – you have met in Section 6.1. The action of fuel cells in these cases is essentially the reverse of electrolysis. In electrolysis, electricity is used to produce hydrogen and oxygen from water. In fuel cells hydrogen and oxygen combine to form water and produce an electric current.

6.4 Considerations

Electrolytic methods require a catalyst embedded in the carbon electrodes to speed up the reaction. PEM cells use catalysts made of platinum and iridium oxide, IrO_2. Both Pt and Ir are expensive and scarce. Ir in particular is critically scarce as it is produced as a by-product of platinum mining. Research is ongoing into transition metal compounds and nano particle platinum as alternatives. Alkaline hydrolysis cells can work with transition metal compound catalysts but are less efficient than PEM cells.

The membrane used for PEM cells is expensive.

Solid oxide electrolytic cells need a high temperature to operate and thus require a source of energy, ideally renewable. These cells need to have steel walls.

Similar considerations apply to fuel cells.

Chemical methods use a metal or porous ceramic membrane to separate hydrogen from carbon dioxide and carbon monoxide. Ceramics in this context are not pottery but oxides or mixed oxides of metals such as aluminium, silicon, titanium and zirconium. The size of the pores needs to be less than 2 nm. Such methods do not produce green hydrogen and need to be combined with carbon capture.

Fermentation is a well-established process and can be adapted to produce hydrogen. This method would potentially reduce the cost of treating biowaste.

Vehicles powered by fuel cells need to have a source of hydrogen in the vehicle. For gaseous hydrogen at 700 bars, a 200 l tank of hydrogen is needed for a car to have a range of 300 miles (similar to that of battery-powered electric vehicles). This is about three to four times the size of petrol tanks. However, the hydrogen tanks can be refuelled from hydrogen supplies at garages in a much shorter time than it would take to recharge a car battery.

Questions

1. Hydrogen-powered vehicles emit water vapour. Although water vapour is a greenhouse gas, this is not a problem. Explain why.
2. What is green hydrogen? Which processes for preparing hydrogen produce green hydrogen?

3. $Be_{12}(OH)_{12}(1,3,5\text{-benzenetribenzoate})_4$ is a MOF which can adsorb 2.3% wt under a pressure of 95 bar, a record for a MOF achieved in 2019. Why does having Be as the metal in this MOF lead to a high %wt adsorption?

4. Write equations for the reactions taking place at the cathode and anode in a solid oxide fuel cell.

Chapter 7

Biomass

Biomass is organic materials such as wood, plants such as corn and soya, straw, and waste. These can be used to make biofuels and electricity. Currently biomass provides the greatest percentage of renewable energy. This chapter covers the various types of feedstock, when and why the energy produced from biomass is regarded as renewable and the effect of biomass harvesting on the environment.

7.1 How Is Energy Produced from Biomass?

Direct methods involve burning biomass to either produce heat or heat water to give steam which then drives a turbine to produce electricity. Steam turbines are used in electricity generators powered by fossil fuels, so these can be modified to use biomass. Combined heat and power plants are where the combustion of fuel to heat water produces steam to drive a gas turbine, and the waste hot water is used to heat buildings. Battersea power station in London was an example of a combined heat and power plant. This power station used coal to produce energy, but from 1950, it used the waste heat to heat homes in Pimlico. Using this type of power plant makes more efficient use of the fuel.

DOI: 10.1201/9781003294337-7

If you have ever tried to light a fire using wet wood, you will have realised that this does not burn well. Biomass contains water, so before being used as a fuel, it must be dried. This can be done by heating it to 200–300°C, a process called torrefaction. The resulting black material is pressed into briquettes.

Alternatively it can be dried by heating at a lower temperature for a longer period of time.

Biomass can also be heated to a high temperature (a process known as pyrolysis) in the absence of oxygen to produce a tar, called pyrolysis oil, plus syngas and biochar, a form of charcoal. Syngas is a mixture of hydrogen and carbon monoxide.

The third use is using biomass to produce gaseous or liquid fuels. Waste, for example land fill, can be broken down by microorganisms in the absence of oxygen. This process produces methane (natural gas). Methane is a powerful greenhouse gas itself but burning it produces carbon dioxide, which is a less powerful greenhouse gas so that using this gas rather than letting it escape to the atmosphere does help reduce global warming. Fermenting biomass gives ethanol. This can be added to petrol or combined with fat to produce biodiesel. Ethanol is widely used in E10 petrol, which is petrol containing 10% of ethanol. Trucks and cars, known as flexible fuel vehicles, are also available that are designed to run on E85 petrol (85% ethanol).

7.2 Why Is Energy Derived from Biomass Considered Renewable?

Biomass in the form of materials such as wood and dung has been burned to provide energy for heating and cooking since prehistoric times and is still used today. However, burning biomass to produce energy releases carbon dioxide along with pollutants such as nitrogen oxides and small particles and is a major cause of respiratory infections due to indoor pollution. Traditional biomass energy like this is not renewable. Using wood to provide energy produces 150% carbon dioxide per unit of energy produced relative to coal and 300–400% relative to natural gas.

Burning biomass to produce energy is only considered renewable if the release of carbon dioxide is balanced by the absorption of carbon dioxide by plants. These plants need not be adjacent to the burner. The aim is to reduce the global concentration of carbon dioxide in the atmosphere. For example, the DRAX power station in the UK is fuelled by wood chips imported from the USA. In 2021, though, DRAX and a French energy generator that also replaced coal with wood were dropped from the S&P Global Clean Energy Index due to uncertainty as to whether these generators combined with carbon capture and storage (Chapter 9) would actually result in net zero carbon dioxide emission.

Plants absorb and use carbon dioxide via photosynthesis. This is a process whereby carbon dioxide and water react to produce glucose and oxygen.

$$6CO_2 + 6H_2O \rightarrow C_6H_{12}O_6 + 6O_2$$

Photosynthesis is a complex process whereby sunlight is absorbed and with the aid of several enzymes produces oxygen gas and protons. Some of the energy from this process is in the form of ATP (adenosine triphosphate), and the protons attached to NADP to form NADPH (nicotinamide adenine dinucleotide phosphate) are then used in the reactions of the Calvin cycle, which uses carbon dioxide to produce glucose (Figure 7.1).

The amount of carbon dioxide captured by plants is known as the net primary production (NPP). This is partly offset by the release of carbon dioxide from decaying plants. The net primary production is influenced by several factors. Drought and higher temperatures reduce the amount of carbon dioxide

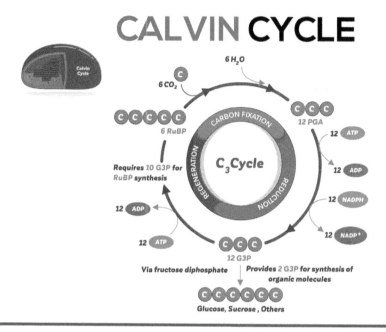

Figure 7.1 **The Calvin cycle.**

absorbed. Soil properties affect the NPP. NPP can be increased by adopting good land management practices. Land management practices that help include planting trees in areas where there were formerly forests, crop rotation, irrigation and change of fertiliser.

7.3 Types of Biomass

7.3.1 Wood

Wood as biomass is used as pellets or wood chips. Pellets are made from material that cannot be sold as timber from a managed forest. Such materials include tree tops, branches, diseased trees and misshapen trees. Waste from sawmills can also be used.

Because trees are generally slow-growing, restoring forests can take a long time to offset carbon emissions emitted by biofuel.

7.3.2 Crops and Grasses

The ten most used crops are switchgrass, wheat, sunflowers, cottonseed oil, jatropha, palm oil, sugar cane, canola and maize. Switchgrass (Figure 7.2a) is a North American prairie

Figure 7.2 **Plants used to provide biofuel: (a) switchgrass, (b)** *Jatropha curcas.*

grass used to feed cattle. Jatropha curcas (Figure 7.2b) is a plant that yields a high percentage of oil and is resistant to drought and pests. It is toxic and mainly used to produce biodiesel. Canola was bred from rapeseed by traditional methods and resembles rapeseed in appearance. It has a high percentage of oil in its seeds.

Crops regrow faster than trees, and so implementation of good farming practices will have a more immediate effect than reforestation.

7.3.3 Algae

Algae are a large group of plants lacking leaves and roots that include seaweed and cyanobacteria. *Spirulina* is a cyanobacterium that has been marketed as a food supplement.

A few species produce poisons. You may have heard of algal bloom (see Figure 7.3). This occurs when the poison-producing algae undergo excessive growth due to factors such

Figure 7.3 **Algal bloom on a small pond.**

as higher temperatures and an excess of nitrogen in the water from fertiliser run-off or sewage.

However, other algae are an attractive proposition for producing energy. One attraction of algae as biomass is that photosynthesis in algae is faster than for other plants. In addition, they can be grown in sea water, do not require soil and take up less space than crops. So far they are not widely used due to their cost.

7.3.4 Waste

Waste such as sewage, dung and chicken litter are generally subjected to pyrolysis or heated to about 700°C with a reduced amount of oxygen to form syngas and slag. Using waste has the advantage that it ensures the waste does not go to landfill where it would decay and emit methane.

7.4 Considerations

To be renewable, producing energy by burning biomass must be offset by the absorption of carbon dioxide by plants. It is also necessary to take into account the energy needed to harvest the biomass, transport it and prepare it for the power plant and whether it is provided by carbon dioxide emitting sources.

The emission of carbon dioxide by biomass burners is instantaneous and continuous. It can take some time for trees to absorb the matching amount of carbon dioxide. Using different models of the type of wood burned, the type of tree planted, the organisation of forests providing the fuel and the energy used to prepare and ship the wood, the time to balance the emissions has been estimated as from 1 year for local forests in Denmark to between 34 and 104 years for the DRAX facility in Yorkshire, England, which imports wood pellets from the USA. Taking into account the energy used to process and transport the biomass, when wood is used as the fuel, the amount of CO_2 emitted per Watt of energy produced is greater than for fossil fuels.

Crops regenerate faster than trees and algae even faster.

Using crops raises concerns that the land used for this purpose reduces that available for food crops. For example, the use of maize to produce biofuels in the USA led to less maize grown for food and consequently a rise in the price of maize food products. The biomass crops also take up water resources.

Grasses such as switchgrass can be grown on land that is unsuitable for food crops.

Continually growing the same crop on an area of land is not environmentally friendly. Also the yield tends to reduce over the years. Good land management, including crop rotation, is preferable.

Even if the emissions and absorptions are balanced, burning biomass also emits pollutants such as nitrogen oxides and small particles which can enter the lungs. The biofuel ethanol is less polluting but does produce carbon dioxide emissions.

Questions

1. A combined power and heat plant in the UK uses wood pellets formed from thinnings, tree tops and other residues in a forest in the USA. List the processes involved in preparing and transporting the forest residues for this purpose. Which of the processes are likely to emit carbon dioxide?

2. The owners of a 20 MW power plant are considering using wheat straw as fuel. This would require 113,000 tonnes of straw per year. The annual yield of straw is 13.5 tonnes per hectare. How many hectares of wheat are needed to provide the necessary amount of straw?

   An average UK farm is 87 hectares.

3. What are the advantages of using switchgrass as a biomass fuel?

Chapter 8

Energy Storage

In this chapter we look at ways of storing energy. Renewable energy sources such as solar cells, wind turbines and tidal generators do not produce electricity continuously. Therefore, it is useful to store excess energy produced when these sources are producing more than needed and release the energy when the sources are not active or providing less energy than needed.

One way to store energy is to use rechargeable batteries. This chapter looks at how batteries work, which types are suitable for storage, the materials needed and current research into alternative types of battery.

Such batteries are also used for transport. While trains and trams can be powered by mains electricity via overhead power lines or rails, cars, lorries, ships and planes need to carry a source of energy with them. Batteries are one way of providing a portable source.

A second storage method is to use capacitors and this is covered briefly.

The third storage method involves using water. In hydro-electric plants, excess energy can be used to pump water uphill into a reservoir which can then feed into the power plant at times of high demand.

DOI: 10.1201/9781003294337-8

8.1 How Rechargeable Batteries Work

Figure 8.1 shows a general electrochemical cell in charging and discharging modes.

In charging mode (Figure 8.1a) the electrons from the external current provider are driven towards the cathode, where reduction occurs. The electrolyte can be liquid, for example in the lead-acid battery found in cars, or solid. When the charged cell is used to power a device or vehicle, electrons are released at the cathode and travel through an external circuit

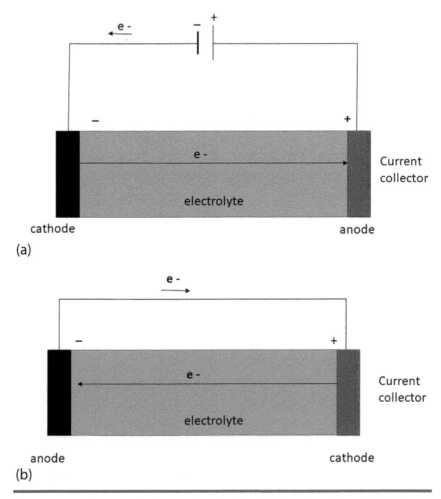

Figure 8.1   A general electrochemical cell (a) under charging (b) in discharge mode.

containing your device or vehicle. You may have noted that when charging your phone, for example, you need a charger and cannot plug it directly into the mains. The charger contains a transformer which converts the AC current from the mains to a DC current. When discharging the battery to power your device, a DC current is produced.

8.2 Batteries for Storage

Batteries can be used to store energy on a wide variety of scales from those in your mobile phone or some hearing aids to large-scale battery complexes feeding the National Grid. When choosing a battery storage system, things to consider include power capacity, energy capacity and cycle lifetime. Power capacity is the power delivered under stated discharge conditions. Energy capacity is the total energy available from the reaction driving the cell. For a particular application, the most suitable battery is chosen using these criteria plus cost, availability of materials and the amount of land needed. Large-scale complexes have used a range of battery types including lead-acid, sodium-sulfur, vanadium redox flow and lithium-ion. Some battery storage facilities use a combination of two types of battery, or a battery bank and another storage method. For example, a storage system in the city of Varel, Niedersachsen, Germany, has two types of storage batteries – lithium-ion batteries which have a high power capacity and sodium-sulfur batteries which have a high energy capacity and can store energy for longer.

8.2.1 *Lead-Acid Batteries*

In lead-acid batteries, the electrolyte is sulfuric acid, the positively charged electrode is lead dioxide and the negatively charged electrode is lead. The equation for discharge is

$$Pb + PbO_2(s) + 2H_2SO_4(aq) \rightarrow 2PbSO_4(s) + 2H_2O(l)$$

Lead-acid batteries are used in cars, including electric cars, to run components such as lights. With older batteries of this type, evaporation of water from the electrolyte meant that you had to top up your battery with water now and then. Valve regulated lead-acid (VRLA) batteries are constructed so that they lose very little water. Lead-acid battery complexes to store energy are in widespread use. One example is a VRLA bank in Springfield, Missouri.

Although lead is an environmental hazard, most lead-acid batteries are recycled.

8.2.2 Lithium-Ion Batteries

In lithium-ion batteries the anode is made of lithium embedded in graphite, forming an intercalation compound, typically C_6Li; in discharge mode, this easily releases Li^+. The anode is coated with copper foil. The lithium ions travel through a Li^+-containing electrolyte to a cathode, where it intercalates. The cathode can be made of mixed metal oxides, e.g. $Ni_xMn_yCo_zO_2$ (Figure 8.2), or other materials that intercalate lithium, notably $FePO_4$, NiO_2 and TiS_2.

The electrolyte is a nonaqueous solvent, such as ethylene carbonate, mixed with a lithium complex salt, lithium hexafluorophosphate ($LiPF_6$), lithium tetrafluoroborate ($LiBF_4$) and lithium triflate ($LiCF_3SO_3$), being commonly used. The Li^+ ions 'rock' between the two intercalation compounds, and no lithium metal is ever present, eliminating many of the hazards associated with early lithium batteries. The cell is encased in aluminium, and a number of cells are linked together and housed in an aluminium and plastic container to form a battery.

8.2.3 Sodium-Based Batteries

In **sodium-sulfur** batteries, the electrolyte is a solid which allows sodium ions to pass through it. Examples are sodium β-alumina and NASICON. Figure 8.3 shows the crystal

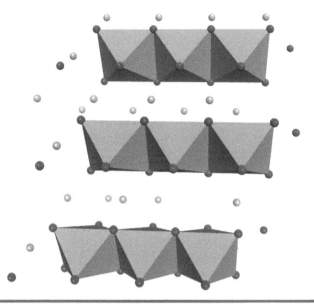

Figure 8.2   The crystal structure of $Ni_xMn_yCo_zO_2$. Intercalated Li^+ ions are shown as small purple spheres.

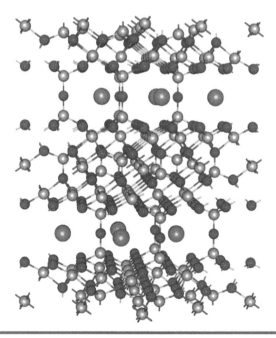

Figure 8.3   Structure of sodium β-alumina. Na large purple spheres, Al small brown spheres, O small red spheres.

structure of sodium β-alumina. This is a layer structure. The sodium ions occupy positions in layers with fewer oxygen ions and vacancies into which the sodium ions can move, thus carrying the current. NASICON, $Na_3Zr_2(PO_4)(SiO_4)_2$ contains a 3d system of channels through which sodium ions can move. One electrode is molten sodium and the other is molten sulfur and graphite. On discharge, the liquid sodium is ionised at the anode to sodium ions and electrons. The electrons enter the external circuit. The sodium ions travel through the solid electrolyte to the cathode where they react with liquid sulfur and electrons from the external circuit to form sodium polysulfides such as Na_2S_5.

The discharge equations can be written as:

Anode reaction: $2Na(l) = 2Na^+ + 2e$
Cathode reaction: $2Na^+ + 2e + 5S(l) = 2Na_2S_5(l)$

Sodium and sulfur are solids at room temperature, and in order to form the molten elements, the battery has to be kept at a temperature of about 300°C. The elements are abundant, but the high temperature means safety precautions have to be taken.

Sodium-ion batteries are similar to lithium-ion batteries as sodium ions can be intercalated in the same way as lithium ions. The anode is carbon, but not graphite as the larger sodium ions need more energy to intercalate into graphite. One possible cathode material is a layered sodium-nickel oxide. The electrolyte contains a sodium compound such as $NaPF_6$. These batteries are already being marketed on a small scale for energy storage. Sodium ions are readily obtainable from sea water, and unlike lithium-ion batteries, the cathode does not contain cobalt. A larger volume is, however, needed to store the same amount of energy as in a lithium-ion battery.

ZEBRA batteries like sodium-sulfur batteries operate at high temperatures. The anode is molten sodium, but the cathode is nickel chloride. The electrolyte is β-alumina (Figure 8.3), but molten $NaAlCl_4$ is inserted between the nickel chloride and the β-alumina to improve the contact between the cathode and the electrolyte.

8.2.4 Redox-Flow Batteries

Redox-flow batteries are rather different from the batteries discussed so far. They consist of two tanks containing electrolyte and metal ions in different oxidation states linked to an array of electrochemical cells. Vanadium flow batteries have been installed in Perth, Scotland, for example, to store power from a farm of solar panels. Figure 8.4 shows schematically a vanadium flow battery in discharge mode. The storage tank feeding into the electrolyte next to the negative electrode contains V^{2+} and V^{3+} ions. The storage tank on the other side contains VO_2^+ and VO^{2+} ions. All the ions are dissolved in sulfuric acid. When the battery is discharging, lilac V^{2+} ions are oxidised to green V^{3+} ions and yellow VO_2^+ ions are reduced to blue VO^{2+} ions. These reactions are reversed on charging.

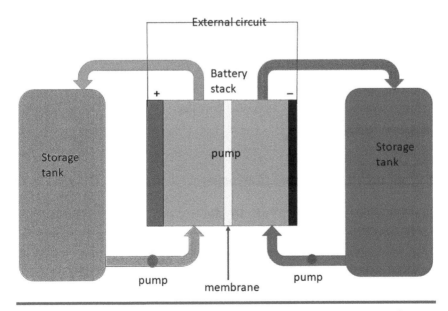

Figure 8.4   **Vanadium flow battery in discharge mode (V^{2+} purple, V^{3+} green, VO^{2+} blue, VO_2^+ yellow).**

The chemical equations for the processes occurring on discharge are

$$VO_2^+(aq) + 2H^+(aq) + e = VO^{2+}(aq) + H_2O(l)$$

and

$$V^{2+}(aq) = V^{3+}(aq) + e$$

The amount of energy stored in such a battery depends on the total amount of the ions taking part in oxidation or reduction present in the storage tanks. The power available depends on the number of electrochemical cells in the battery stack.

8.3 Batteries for Transport

Lead-acid batteries have been used to power milk floats and are still used for forklift trucks in warehouses. Although very early electric cars also used lead-acid cells, these are heavy and the speed and range of the cars were limited. Thus electric cars have only become popular with the development of lithium-based batteries. This section therefore describes the lithium-based batteries used now, plus research into improved lithium-based batteries and sodium-based batteries that could be used in the future.

8.3.1 Lithium-Based Batteries

The batteries used currently in electric cars are lithium-ion batteries. The battery used in the second generation Nissan Leaf electric car, for example, has a layered cathode consisting of layers of oxide ions, $LiCo_xMn_yNi_zO_2$, as shown in Figure 8.3.

These batteries are similar to those in your phone, but the car has not just one but a bank of batteries. Car manufacturers guarantee their batteries for about 8 years, but in the future they may last 20 years or more. Several schemes use batteries that are no longer working efficiently enough to power cars as small-scale storage. Eventually, the batteries can be taken apart and components recycled. The lithium-ion batteries we described in the previous section have an electrolyte that is a liquid solution of a lithium salt. Researchers are looking at replacing the solvent with a solid that conducts lithium ions. Examples are structures based on a form of lithium phosphate and garnets containing excess lithium ions. An alternative is the lithium-ion polymer battery where the solvent is replaced by a conducting polymer gel.

Cobalt is predominantly mined in the Democratic Republic of the Congo. Conditions in these mines have been condemned on human rights grounds. Researchers are looking into cobalt-free cathodes, initially by replacing cobalt with more nickel but also replacing cobalt and nickel with the more abundant and sustainable iron.

Researchers are also looking at replacing lithium with the more available metals, sodium and calcium.

8.4 Capacitors

Capacitors are widely used in electronic circuits. They consist of two conducting plates separated by an insulating dielectric. When a current is applied to the plates, they acquire an electrical charge. The dielectric prevents the positive charges on one plate from combining with the negative charge on the other. These charges remain until the capacitor is discharged through an electrical circuit. Capacitors used for large-scale energy storage are electrochemical or double layer capacitors. These have carbon electrodes attached to the metal plates and are separated by an electrolyte. The carbon electrodes are typically nanoporous to provide a large surface area for the charge to form on. The electrolyte can be strongly acidic or alkaline aqueous solutions or organic salt solutions. Charge builds up at the carbon/electrolyte interface with negative charges on the carbon and positive charges on the electrolyte adjacent to the carbon. Unlike batteries, no chemical reaction occurs. Because of this, discharge is virtually instantaneous and for this reason such capacitors are used with uninterruptable power supplies such as on airplanes. They are also used in regenerative braking systems in modern cars.

8.5 Pumped Storage Hydropower

In pumped storage hydropower, water is pumped up to a reservoir where it is stored until needed. The water then returns to a lower reservoir producing electricity via a water turbine in the same way as hydroelectric power. Such systems have been around since the late nineteenth century and are used worldwide.

Questions

1. Fill in the cathode material, anode material and electrolyte for the storage batteries in the table below.

Battery	Cathode material	Electrolyte	Anode material
Lead-acid			
Lithium-ion			
Sodium-sulfur			
ZEBRA			

2. What would happen if you try to charge a battery directly from the mains?
3. Why did the development of lithium-containing batteries make cars powered by batteries a commercial proposition?
4. How do capacitors differ from batteries?

Chapter 9

Carbon Capture, Storage and Conversion

The aim of renewable energy systems is to reduce the amount of carbon dioxide being released into the atmosphere, but the replacement of fossil fuel energy sources will not be sufficient to reduce the amount of carbon dioxide in the atmosphere to an acceptable level over the next 20 years. One way to tackle this is to remove carbon dioxide from the atmosphere. How can this be achieved? How long will the storage last before the carbon dioxide is re-released? Having removed some of the carbon dioxide, can we use it? This chapter looks at ways of capturing and storing carbon dioxide and reacting it to make useful products.

9.1 Carbon Capture and Storage

9.1.1 Trees and Peat Bogs

Plants absorb and use carbon dioxide by photosynthesis. This is a process whereby carbon dioxide and water are reacted to produce glucose and oxygen.

DOI: 10.1201/9781003294337-9

$$6CO_2 + 6H_2O \rightarrow C_6H_{12}O_6 + 6O_2$$

This reaction is thermodynamically unfavourable. Under normal conditions, the favoured reaction is that of glucose reacting with oxygen to form carbon dioxide and water. This reaction is, luckily, slow with a high activation energy, so glucose does not spontaneously react with oxygen in the air. For the formation of glucose to occur, CO_2 and/or H_2O have to react to form higher energy molecules. In plants there are two connected processes. The first is splitting water to produce oxygen, protons and electrons.

$$2H_2O \rightarrow O_2 + 4H^+ + 4e$$

This occurs via the absorption of sunlight. In addition, the rate of reaction is increased by lowering the activation energy using catalysts, in this case enzymes.

The second process is carbon dioxide's reaction with hydrogen. Carbon dioxide is processed in the Calvin cycle (Chapter 7, Section 7.2) to give glyceraldehyde-3-phophate, which is used to synthesise glucose.

This is a complex process, but clues as to how to approach chemical conversion of carbon dioxide can be taken from the approach.

1. An input of energy is needed.
2. Catalysts are needed. Enzymes used in the oxidation of water use Mn and Ca. An enzyme involved in the reduction of NADP contains Fe-S clusters. This gives a clue as to metals that might be used.
3. Splitting of water (here to form H^+ + O_2 but could be electrolysis of water in industrial processes) and reduction of carbon dioxide occur as two separate processes.

Trees can have lifetimes of hundreds or even thousands of years, and the carbon dioxide they use is locked up for the lifetime of the tree. Peat forms in wet, acidic conditions as layers of partly decomposed moss and other plants. To make a 1 m thick layer of peat can take a thousand years. The importance of peat lies in the wet, acidic conditions preventing carbon in the peat from being released into the atmosphere.

9.1.2 Construction Materials and Minerals

Concrete is the second most widely used material after water. Traditional concrete is made by binding aggregates such as sand, gravel or crushed stone with a cement–water paste. Cement is currently responsible for a significant emission of carbon dioxide (approximately 8% of global emissions}. The widely used Portland cement can be made by heating limestone (calcium carbonate) with clay (layered aluminosilicates), resulting in the formation of calcium silicate, calcium aluminate and carbon dioxide.

Concrete can last for 50–100 years, and over this period it absorbs 10–30% of the carbon dioxide emitted in its manufacture. Replacement materials or changes in production methods can reduce carbon dioxide emissions. When disposed of, fresh surfaces are exposed, enabling additional absorption of carbon dioxide. Aggregates can be recycled. Flexible or bendable concrete is conventional concrete that contains fibres. Not only is this concrete more flexible than traditional concrete, but it is also self-mending giving it a longer lifetime.

9.1.3 Porous Solids

Porous solids have structures containing cages of channels which can accommodate small molecules. Zeolites are one example of these solids. Figure 9.1 shows the cages and channels in the structure of a well-known zeolite, ZSM-5. Naturally occurring zeolites contain water in their pores and when heated with a blow torch, bubble as though they were alive as the water is driven off.

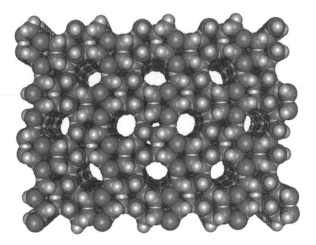

Figure 9.1   Computer model of the zeolite ZSM-5.

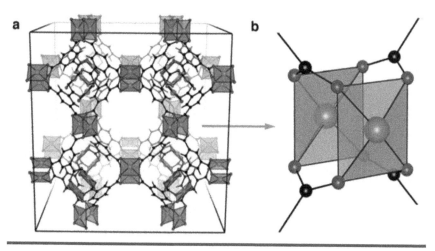

Figure 9.2   (a) Crystal structure of anhydrous HKUST-1. (b) The Cu$_2$ 'paddle wheel'. (Figure 1 in Christopher H. Hendon and Aron Walsh, *Chem Sci*, 2015, 6, 3674, This article is licensed under a Creative Commons Attribution 3.0 Unported Licence.)

Others include metal organic frameworks (MOFs) and covalent organic frameworks (COFs). Metal organic frameworks consist of metal clusters linked by organic ligands. Figure 9.2 shows a portion of the crystal structure of a MOF,

Mg-HKUST-1. The metal cluster is a Cu_2 'paddle wheel' as shown in Figure 9.2b.

The linkers in this MOF are 1,3,5-benzene-tricarboxylate.

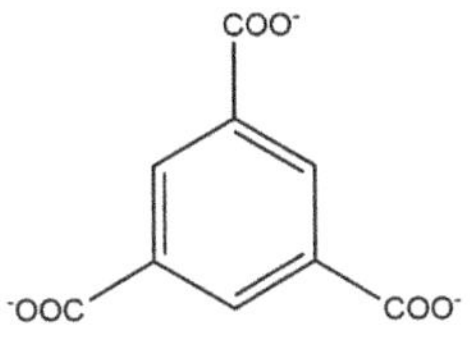

Pore size can be tailored by altering the type and length of the ligand. To capture carbon dioxide from the air, the pores have to accommodate CO_2, in competition with other small molecules in the air such as nitrogen and oxygen. Some MOFs with very high specificity for carbon dioxide have been synthesised.

9.1.4 Absorbance by Liquids

Carbon dioxide can be absorbed by slurries of alkalies such as MgOH and CaOH to give the carbonate MCO_3 as a precipitate. These carbonates can then be used to make a variety of products, as you will see in Section 9.2.

An industrial process for removing carbon dioxide from flue gases uses 2-aminoethanol NH_2CH_2CHOH. The flue gases are first treated with water to remove water-soluble components, and sulfur dioxide is removed by the addition of soda ash (sodium carbonate). The remaining gas is reacted with a concentrated solution of 2-aminoethanol.

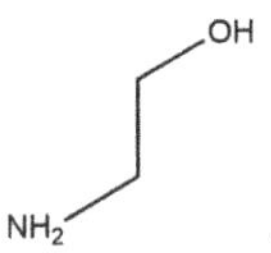

2-Aminoethanol reacts with carbon dioxide to form a carbamate

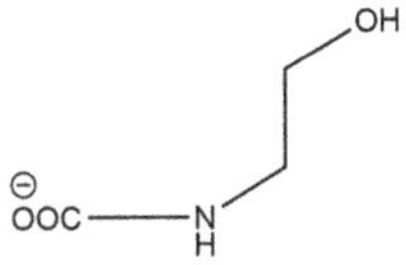

This reaction is reversible, and the carbamate is heated to reform carbon dioxide, which is used to make fizzy drinks. It can also be used to manufacture chemicals such as sodium bicarbonate.

Once captured, carbon dioxide can be stored in places such as underground caverns and disused oil and gas fields. However, this is an expensive process, although the carbon dioxide remains stored for many years. Rather than just storing captured carbon dioxide, why not use it to manufacture chemicals that we now obtain from oil or other carbon sources? The next section explores progress in this field.

9.2 Conversion of Captured Carbon Dioxide

9.2.1 Mineralisation

Silicate minerals, in particular those containing magnesium such as olivine (Figure 9.3), will react with carbon dioxide to produce carbonates.

Magnesium carbonate has many uses, for example, as an antacid, in cosmetics, as a drying agent and in flooring. It is also used to manufacture magnesium oxide. The last use releases the captured carbon dioxide, but because the carbon dioxide was captured from the air, it is better than using minerals such as dolomite (calcium magnesium carbonate). In other cases the carbon dioxide is retained.

The use of naturally occurring minerals involves mining and transport, processes which emit carbon dioxide.

An alternative is to use alkali waste materials containing calcium and/or magnesium to absorb carbon dioxide. Treatment of industrial alkali waste with carbon dioxide can give useful products, such as aggregates, to use in construction. This avoids the emissions associated with the extraction and transport of aggregates. For example, carbon dioxide emitted by steel works can be reacted with the slag produced in the manufacture of steel. Other waste sources that can absorb carbon dioxide are fly ash and concrete waste.

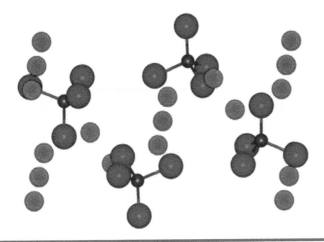

Figure 9.3 **Crystal structure of olivine. Green spheres are Mg and Fe, red spheres oxygen and small grey spheres silicon.**

9.2.2 *Chemicals and Fuels*

You saw above that trees used light and catalysts (enzymes) to make glucose. That is, an energy input and a catalyst are required. Similar techniques can be used to react carbon dioxide to give products such as formaldehyde and methanol. Methanol is widely used as a solvent and as a precursor for other chemicals. The products from the reduction of carbon dioxide can be used to make polymers.

Thermochemical methods use heat, high pressure and a catalyst to react CO_2 with a reducing agent such as hydrogen gas. Examples are the production of methanol from carbon dioxide and hydrogen gas and the production of carbamates from carbon dioxide and urea.

In electrochemical catalysis, an aqueous solution of CO_2 is electrolysed. Oxygen is produced at the anode, and CO_2 is reduced at the cathode. This method would be expected to use excess energy from renewable sources. In aqueous solutions at room temperature and pressure, Cu electrodes produce methane, CH_4, and a mixture of C_2 molecules such as ethane, C_2H_6. Ag, Au, Zn and Pd typically produce CO, and Sn, In, Pd and Bi produce formate, HCOO⁻. Higher pressures and the use of mixed ionic solvents or aprotic solvents are being investigated. Aprotic solvents such as dimethylsulfoxide (DMSO) reduce the probability of the competing reaction of hydrogen production at the cathode.

In the photochemical process, a catalyst is suspended in the carbon dioxide solution and sunlight shines on the solution. Carbon dioxide is reduced to either carbon monoxide and hydroxide ions or formate ions.

In the photoelectrochemical process, the carbon dioxide is reduced by electrons from a semiconductor/electrolyte interface or via a redox system.

As well as chemical methods, biological methods of carbon dioxide conversion have attracted interest. Several research projects are looking at ways of using genetically modified microorganisms such as bacteria to produce biofuels or proteins. The microorganisms can produce compounds such as acetic acid and alcohol from carbon dioxide and hydrogen.

Thermochemical and electrochemical processes are used commercially.

If carbon dioxide is used to make fuels, then some will be released when the fuel is burned. Analysis of the impact of such methods therefore has to take into account the amount released and the time span over which it is released. A similar analysis for chemicals has to consider the lifetime of the product and what happens when it is disposed of.

Questions

1. Although carbon dioxide is a major contributor to global warming, it is only a very small fraction of the composition of our atmosphere. Suggest ways in which carbon

dioxide could be captured from the air separately from the major components of the atmosphere.
2. List the ways in which carbon dioxide can be stored.
3. Where is energy input required in the thermochemical reaction converting carbon dioxide to methanol?

Index

9 781032 275758